Hot Zone Log:
Personal HazMat Record

LTC Stanley E. Toy

THOMSON
DELMAR LEARNING™

Australia Canada Mexico Singapore Spain United Kingdom United States

THOMSON
DELMAR LEARNING

Hot Zone Log: Personal HazMat Record

LTC Stanley E. Toy

Vice President, Technology and Trades SBU:
Alar Elken

Editorial Director:
Sandy Clark

Acquisitions Editor:
Alison Weintraub

Development:
Stacey Wiktorek

Marketing Director:
David Garza

Channel Manager:
Bill Lawrensen

Marketing Coordinator:
Mark Pierro

Production Director:
Mary Ellen Black

Production Editor:
Toni Hansen

Editorial Assistant:
Stacey Wiktorek

Printed in the United States of America
1 2 3 4 5 XX 06 05 04

For more information contact
Thomson Delmar Learning
Executive Woods
5 Maxwell Drive, PO Box 8007,
Clifton Park, NY 12065-8007
Or find us on the World Wide Web at
www.delmarlearning.com

Library of Congress Cataloging-in-Publication Data:
Card Number:

ISBN: 1-4018-7229-8

NOTICE TO THE READER

Publisher does not warrant or guarantee any of the products described herein or perform any independent analysis in connection with any of the product information contained herein. Publisher does not assume, and expressly disclaims, any obligation to obtain and include information other than that provided to it by the manufacturer.

The reader is expressly warned to consider and adopt all safety precautions that might be indicated by the activities herein and to avoid all potential hazards. By following the instructions contained herein, the reader willingly assumes all risks in connection with such instructions.

The publisher makes no representation or warranties of any kind, including but not limited to, the warranties of fitness for particular purpose or merchantability, nor are any such representations implied with respect to the material set forth herein, and the publisher takes no responsibility with respect to such material. The publisher shall not be liable for any special, consequential, or exemplary damages resulting, in whole or part, from the readers' use of, or reliance upon, this material.

Table of Contents

Preface

The *Hot Zone Log* is intended to assist personnel assigned within the hazardous material response profession to better track individual training and actual response performance. Modeled after a standard dive log, this record serves as a personal historical document of operations conducted in the hot zone while wearing personal protective equipment or PPE, more specifically level-A or level-B and the use of self contained breathing apparatus (SCBA) equipment. Much of the information provided in this log serves as tools to assist in the operational management and considerations associated with a hazardous materials (HazMat) incident. Designed specifically to address safe practices and procedures, this log also provides examples of proven emergency considerations that are essential to conducting safe operations in the hot zone.

This log, if properly used by the individual, will serve to chronologically document essential elements to actual response and or training. This hot zone log is designed to effectively record the number of incidents one has responded to; the degree of hazardous products one has been exposed to; completion of performance standards; decontamination considerations; and pre and post entry medical monitoring status. An accurate accounting of these elements will serve to offer a quick historical reference to individual HazMat response experiences. Additionally, this log will complement any organizational health and safety program as well as enhance existing risk management policies. This log is not intended to supersede any existing organizational requirements, regulations, or standard operating procedures (SOP). All of the procedural considerations provided in this log are "recommendations" and not specified standards by law. The hot zone log is a means to efficiently and effectively address individual performance first and foremost. The aim is to instill consistent record keeping and monitoring for those who have been designated to "*enter the hot zone*"!

The author and staff at Thomson Delmar Learning gratefully acknowledge the time and effort of those who reviewed this project. The comments and suggestions of the review panel were invaluable. Our sincere thanks to:

Jack D. Lee Jr., Assistant Chief, Special Operations, Federal Fire Department, Honolulu, Hawaii

Captain Dave Matthew, Wichita-Sedgwick County HazMat Team, Kansas

Meridith McConnell, Firefighter/Instructor, Granite Shoals Fire Department, Texas

Corey Molinelli, HazMat/WMD Specialist, Deltona, Florida

Don Pulliam, Counter-Terrorism Consultant, DC Environmental, LLC, Port Vincent, Louisiana

Lt. William Tatsch, Training Officer and HazMat Response Coordinator, Granite Shoals, Texas

Brad Morell, Booz Allen Hamilton, Abingdon, Maryland

Kevin Fogarty, ATEC Consultant, Boynton Beach, Florida

SECTION 1

Personal Information

Name/Rank: ______________________

DOB: ______________________

Organization or Unit of Assignment: ______________________

Alarm Box Number: ______________________

Business Telephone Number: ______________________

Cell Phone: ______________________

Emergency Contact Information

Name of Contact: ______________________

Relationship: ______________________

Address: ______________________

Phone: ______________________

Cell Phone: ______________________

Medical Information

Blood Type: ______________________

Allergies: ______________________

Medication: ______________________

When Taken: ______________________

Height: ______________________

Average Weight: ______________________

Medical Insurance

Company/Organization: ______________________

Policy Number: ______________________

Doctor's Name: ______________________

Phone Number: ______________________

SECTION 2

OSHA Safety Standards and Practices

The office of Occupational Safety and Health Administration (OSHA) has defined **HazMat** as those hazardous materials or hazardous products that pose a potential health risk to humans or the environment.

CAUTION: HazMat technicians tasked to "safely" mitigate hazardous materials must have knowledge of the potential chemical, biological, or radiological properties of hazardous products. These properties include *flammability*, *toxicity*, *reactivity*, and *energy sensitivity*. Additionally, HazMat technicians should have a working knowledge of the physical properties of hazardous products. The physical properties of HazMat include *state of form*, *solubility*, *state of changes*, and *density*, as well as other variables that may influence a product's physical state.

For hazardous materials technicians and specialists, the federal regulation of the OSHA safety standards and practices requires that appropriate HazMat training be completed to address the care, use, and/or testing of chemical protective clothing, including the use of fully encapsulating personal protective suits, use of SCBA, application of a medical surveillance program, and planning the execution of offensive operations to be conducted in an established hot zone where the mitigation of hazardous products that may have spilled or leaked from their container is accomplished (29 CFR Hazardous Materials 1910.120 Appendix C).

HazMat Chiefs or Captains may be expected to support and advise an on-scene incident commander (IC) as to what actions should be taken to bring about the mitigation of a hazardous environment. Typically the most experienced HazMat personnel on-scene is traditionally assigned this task. These personnel must be fully knowledgeable about the proper response protocols and procedures specific to their organization. They should also serve to assist in the execution of tactical goals and objectives as defined by the IC. Additionally, these individuals may be required to know where and how to obtain additional assistance and to be familiar with the local district's emergency response plan and the state emergency response plan.

NOTE: There are two National Fire Protection Association standards: NFPA 472—"Standard for Professional Competence of Responders to

Hazardous Material Incidents" and NFPA 471 "Recommended Practice for Responding to Hazardous Material Incidents," which are excellent resource documents to aid fire departments and other emergency response personnel in developing their training programs. NFPA 472 provides guidance on the skills and knowledge needed for first responder awareness level, first responder operations level, HazMat Technicians, and HazMat Specialist. NFPA 472 also offers guidance to personnel who may be charged with managing the tactical and strategic challenges of hazardous materials incidents.

CAUTION: The NFPA standards are not law; however, the standards offer proven practices and procedures as defined by the industry and profession of responders that certainly complement organizational SOP. *Reference:* www.osha.gov.

The skill sets and training guidelines provided serve only as a base for HazMat response personnel who may have to operate in a designated hot zone, a task usually given to HazMat Technicians. By definition, **HazMat Technicians** are those immediate response hazardous materials personnel who are trained, equipped, and tasked to perform direct offensive operations in a hot zone; such as spill or leak control and other mitigating activities as necessary to render safe a hazardous materials incident. The training guidelines are traditionally applicable to HazMat response units; however, explosive ordinance detachments (EOD), weapons of mass destruction civil support teams (WMD-CST), or Department of Health environmental-health and radiological response teams are certainly included in the community of trained HazMat responders.

NOTE: These guidelines are not intended to supersede prescribed departmental training requirements but rather to enhance existing programs.

SECTION 3

Isolation Zones and Personal Protective Posture

Figure 3-1 serves to illustrate one example of establishing hazard zones as part of developing an incident action plan. **Zones** are operational areas, established at HazMat incidents, within which only specific types of emergency response operations are permitted. Personnel working in these areas must follow organizational SOP to ensure operations are conducted safely and efficiently. The size of each zone is determined by the initial hazard assessment commonly referred to as the *size up*. *(Reference: Operational Training Guidance, Civil Support Teams NGB - 2000.)*

The physical location of each zone is influenced by wind direction, weather, terrain, and manmade structures. Defining the location of the zones early in a HazMat incident will facilitate rapid placement of supporting response capabilities.

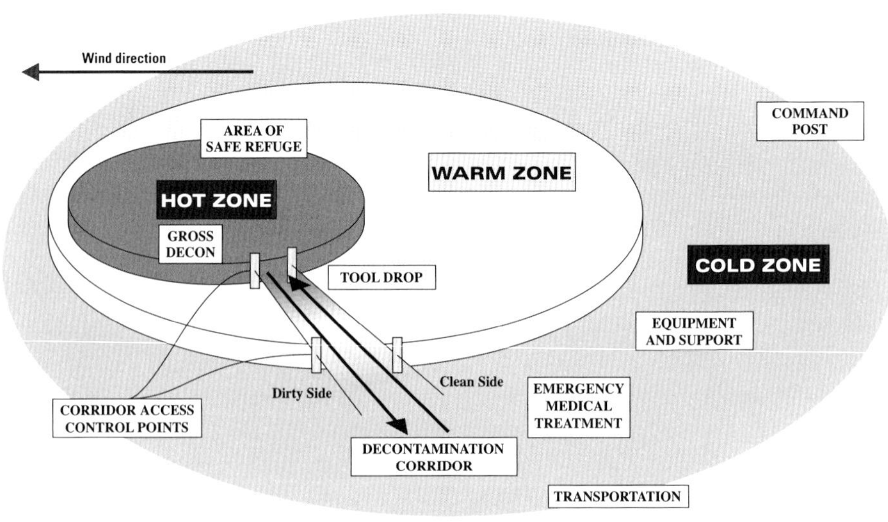

Figure 3-1

CAUTION: An *all hazards* environment can potentially have the presence of products that are Chemical, Biological, Radiological, Nuclear, and or Explosive (CBRNE) presenting an extremely volatile environment in which to conduct operations.

For response to an all hazards environment or where the identification of hazardous products is not known, always respond with the highest level of protection. It is recommended that the designated safety officer refer to the NIOSH Pocket Guide and PPE Compatibilities Index to best determine the chemical permeation resistance of a responder chemical suit before responding.

To determine the most appropriate level of PPE required in response to a HazMat incident, one must obtain information on the substance involved, concentration/quantity, and relative information on the physical environment.

NOTE: There are uses and environments and chemicals for which HazMat protective suits and/or fabrics are unsuitable. It is the responsibility of the responder and safety officer to review available data and verify that the HazMat protective suits and/or fabrics are appropriate for the intended use and meet all specified government and industry standards. CAUTION: Do not use HazMat protective suits for fire protection. Avoid open flames and heat. For technical assistance call:

Technical Services Document and Chemical Permeation Resistance

DuPont-Kappler Technical Services Department Toll Free 1-800-633-2410

Isolation Zone Category	Responder Activity Actions	Personnel Training	Personal Protective Equipment (PPE)
HOT ZONE The isolation zone where the highest presence or potential presence of contaminated exposure exists warrants access by those personnel trained to perform specific operations.	**Survey/Sampling** Operations and actions conducted by response personnel include hazard detection and identification, area and point source monitoring	■ HazMat Technicians ■ Chem-Bio-Radiological and Explosives Specialists ■ Urban Search and Rescue	**Level A** ■ Fully encapsulated HazMat suit ■ Self Contained Breathing Apparatus ■ Chem resistant boots and gloves
WARM ZONE The isolation zone where decontamination operations occur serves as a buffer or safety zone between hot zone and cold zone, typically upwind and on level terrain from hot zone.	**Decontamination Operations** To provide technical or gross decontamination functions for responder personnel exiting hot zone. Type of decontamination operations are influenced by type of contaminants personnel have been exposed to.	■ Decontamination Specialist ■ HazMat Technicians ■ Medical Response Personnel ■ Hazard ID Specialist/Analysts	**Level-B** ■ Semi encapsulated suit ■ Self Contained Breathing Apparatus ■ Chem resistant boots and gloves
COLD ZONE The isolation zone considered safe. Upwind of warm and hot zones	**Support Operations** All the responder functions occur to include the location of the incident command post, emergency medical treatment and holding areas, communications, transportations and logistical support personnel.	■ On-Scene Incident Commanders ■ EMS Personnel ■ Safety Officers ■ HazMat Operations Personnel ■ Communications Personnel ■ Others	**Level C and Level D** ■ In the event of patient contact Level C for Patient Care to include TYVEX over garment ■ Face shield, safety glasses ■ Hard hat ■ Splash-resistant clothing ■ Steel toed boots if structural hazards are present

Table 3-1 Isolation zone and personal protective posture matrix *(Reference: CDC December 2001)*

SECTION 4

Risk Management Considerations

By definition, Risk Management is a five-step process of *identifying, assessing*, and *controlling* risks arising from operational factors and making decisions that balance risk cost with mission benefits. Risk management supports and encourages initiatives allowing for flexibility, adaptability, and eagerness to act that enhance a safer operational environment.

The five-step risk management is a process of identifying and controlling hazards to conserve personnel and resources:

Step 1: Identify the hazards.
Step 2: Assess the hazards to determine the risks.
Step 3: Develop control measures and make risk decisions.
Step 4: Implement control measures.
Step 5: Supervise and evaluate risk management measures.

Conducting operations in a hazardous environment is both physically and mentally demanding. Having proper training and well-maintained equipment and understanding the potential for an all hazards environment are just the basic requirements necessary to conduct operations in the hot zone. A significant consideration that is relative to today's response, especially after the terror attacks of September 11, 2001, is the threat of terrorism with the potential use of weapons of mass destruction. Commercial off-the-shelf toxic industrial agents can be readily obtained and effectively used as weapons by a would-be terrorist. This added threat presents an asymmetrical challenge to responders who are tasked to *manage, define, identify*, and *mitigate* a crisis of this magnitude.

One of the key staff personnel tasked to initiate risk management procedures is the on-scene *safety officer*. The **safety officer** should continually update the on-scene incident *site safety health plan* to address the associated hazards at the incident site. The **site safety health plan** is a stand-alone document that enables safety officers to address any and all potential safety concerns that may be impacted by the incident response. Staff members must inform the IC of recommended control measures to reduce the level of risk concerning what may be identified as hazards in the site safety health plan.

Hazards and the resulting risks may vary as circumstances and the operational focus of the incident change.

NOTE: All personnel supporting an on-scene IC must develop a sense of awareness for ever-changing hazards such as those associated with the environment (weather; visibility, contaminated air, water, and soil), equipment readiness, individual and unit experience, and fatigue. Leaders should always keep their immediate chain of command informed of the on-scene risks and changing conditions that may influence risk reduction efforts. By incorporating risk management measures in the operational plans at the incident site, IC are better able to ensure the operational safety of all assets and personnel providing response support at a HazMat incident.

Reference: DOD Field Manual 100-14

SECTION 5

Hot Zone Hand and Arm Signals

The following photographic illustrations are provided to offer an example of how an individual dressed out in level-A PPE would render hand and arm signals as a way to communicate in the event there is a loss or lack of radio communications. Clearly normal physical functions are limited when operating in a level-A suit, so it is important that all signals given are exaggerated and deliberate for their desired intent to be understood. The inclusion of these signals can serve as a backup to the primary means of radio communication. Additionally, each of these signals is designed to enable responders to effectively communicate specific actions or tasks that may be accomplished without relying on two-way radio communication.

Basic Hand and Arm Signals

NOTE: UNDERLINED SIGNALS ARE A MUST TO REVIEW PRIOR TO ENTRY

Numbers: for 1–5, use one hand, hold palm up and use the appropriate number of fingers to indicate the number (figure 5-1). For numbers 6–9, hold the arm parallel to the ground and extend the appropriate number of fingers across the chest (figure 5-2). For the number 0, simply hold a closed fist. This technique takes a little bit of practice, but allows only one hand to be used. To stop between numbers drop arm down at side.

Figure 5-1

Figure 5-2

Communications check: make a closed fist. Extend the thumb and pinkie and place the pinkie in the vicinity of your mouth.

Figure 5-3

Negative (to indicate a negative response): clench fist, extend thumb and point thumb toward ground.

Figure 5-4

Air check: hold hand open as if holding a big sandwich (or putting a regulator into your mouth). Move the hand toward your mouth. The response can either be a thumbs up/down, or by indicating a single number followed by a short break, followed by a second number. For example, to indicate that you have half a tank of air, indicate the number 1, wait a second and then indicate the number 2.

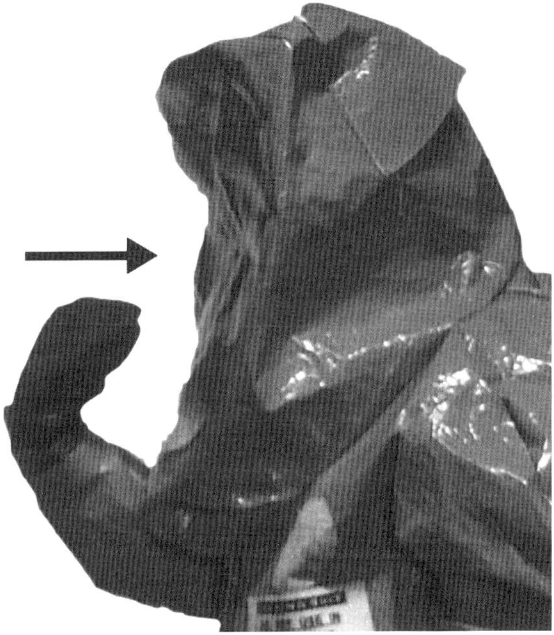

Figure 5-5

Pace count: tap the thigh with one hand.

Figure 5-6

Map check: place both palms together in front of your chest with fingers pointing away from your body. Open your hands at the top until both hands are parallel with the ground and the pinkies are touching.

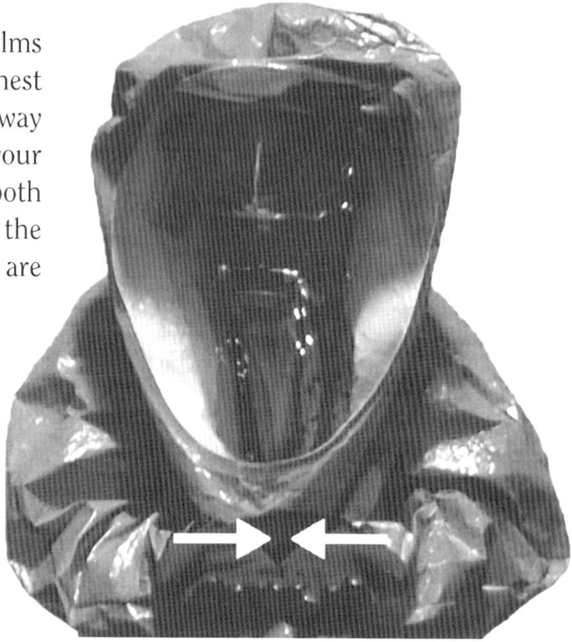

Figure 5-7

Picture (to tell a team member that you want him to take a picture of something): simply point to the person you want to take a picture, pretend you are taking a picture, and then point to the object you want photographed.

Figure 5-8

Tripwire: touch the trigger finger to the thumb with both hands, slowly pull the hands away from each other, and then point to where the tripwire is.

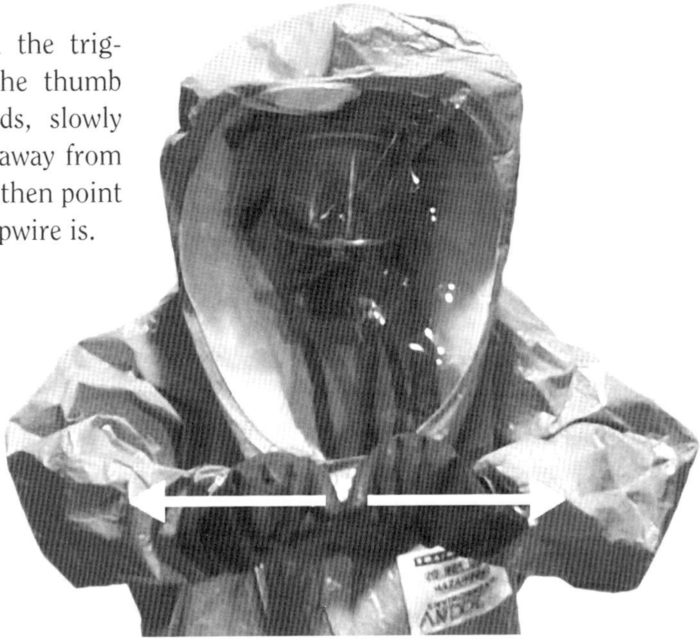

Figure 5-9

Abort: flatten hand, palm side down and slash hand across throat.

Figure 5-10

Danger (to indicate that an object is dangerous): make a fist, and then extend the trigger finger and the thumb. Point the thumb toward the ground and the trigger finger toward the dangerous object.

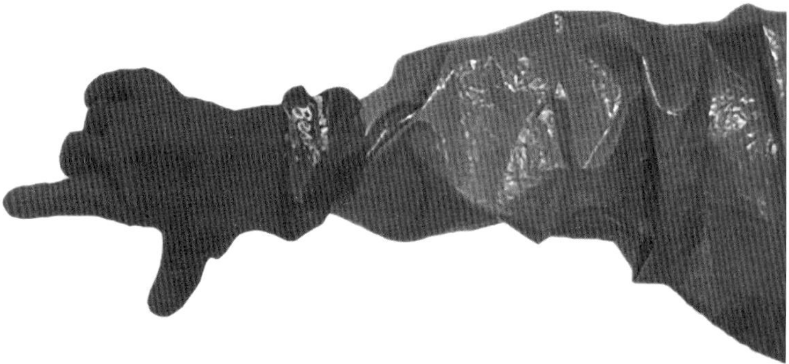

Figure 5-11

No air: choking sign.

Figure 5-12

Trouble (to indicate you are experiencing trouble): wave hands above head.

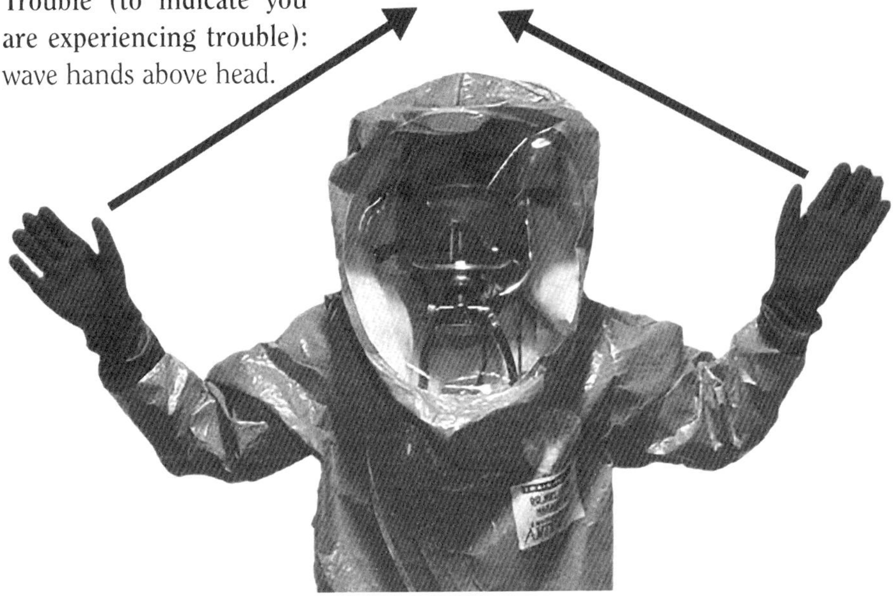

Figure 5-13

Positive (to indicate "yes" or "affirmative" or "everything is good"): thumbs up.

Figure 5-14

Rally (to indicate your desire to have the team close on you): wave one hand above head.

Figure 5-15

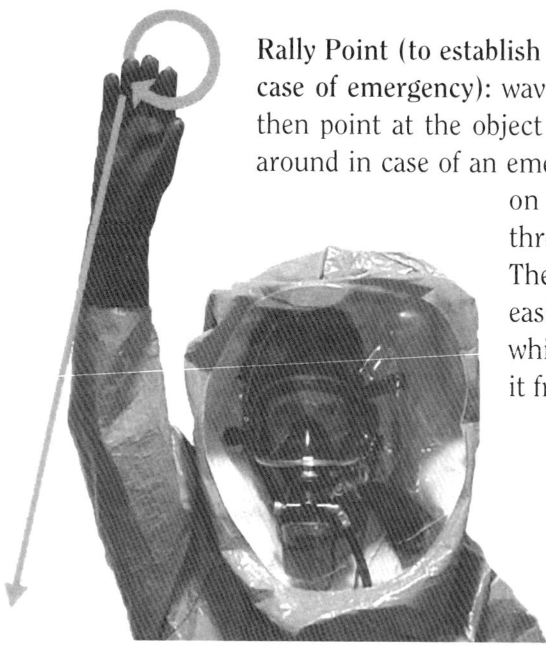

Rally Point (to establish a point on which to rally in case of emergency): wave one hand above head and then point at the object you want everyone to rally around in case of an emergency. Personnel will pass on the signal as they pass through the rally point. Note: The object must be unique and easily identifiable regardless of which direction you approach it from.

Figure 5-16

Time (primarily to indicate that "it's time to go," may also be used to ask how long an event has taken to complete): slap wrist (tap at watch).

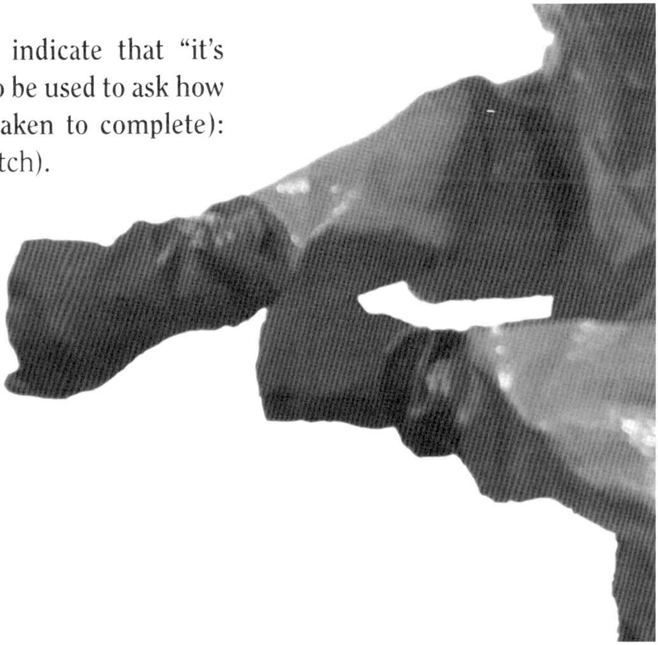

Figure 5-17

Look at something: point two fingers at eyes, then at object.

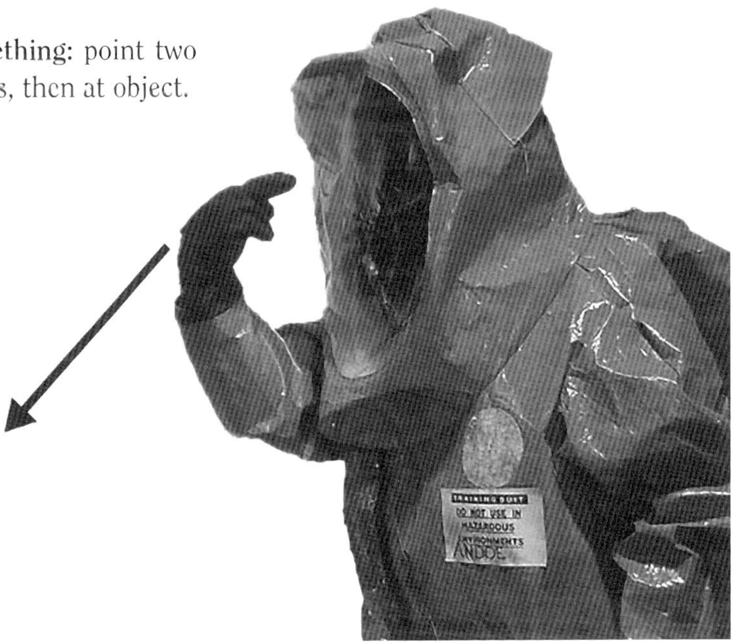

Figure 5-18

I can't hear: point to ear.

Figure 5-19

I can't talk (my radio isn't transmitting): open and close fingers to thumb. Movement is toward side of head or towards ear.

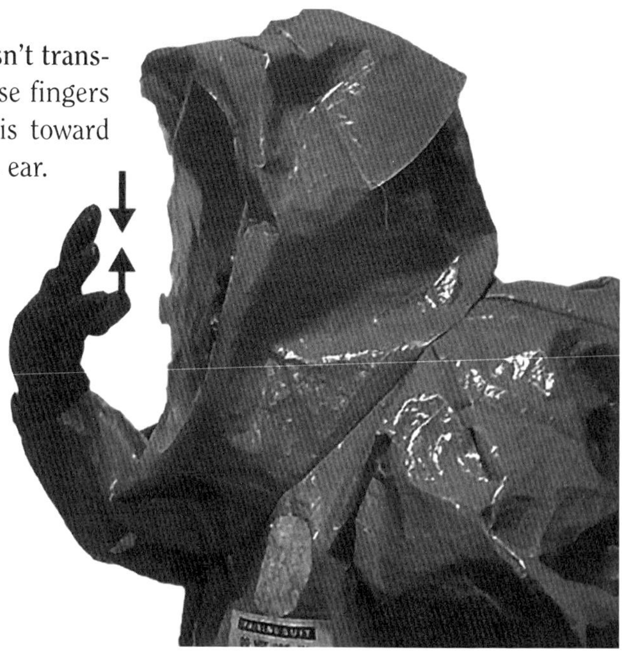

Figure 5-20

I don't understand/ What?: raise both hands to your side with the palm up.

Figure 5-21

Cut me out of suit: slash one hand across body from shoulder to waist.

Figure 5-22

Halt/ Stop: open hand and raise with palm facing in same direction as head.

Figure 5-23

Sample: make a fist or shape hand as if holding a tool, and then simulate movement of "scooping".

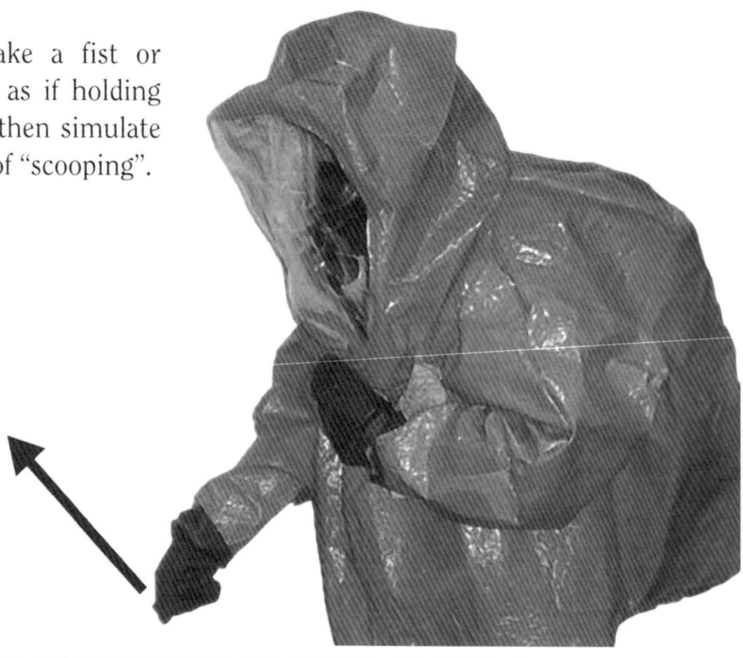

Figure 5-24

Come/come to me: open hand, aim hand towards the person you wish to signal, and then bring hand and arm toward your chest.

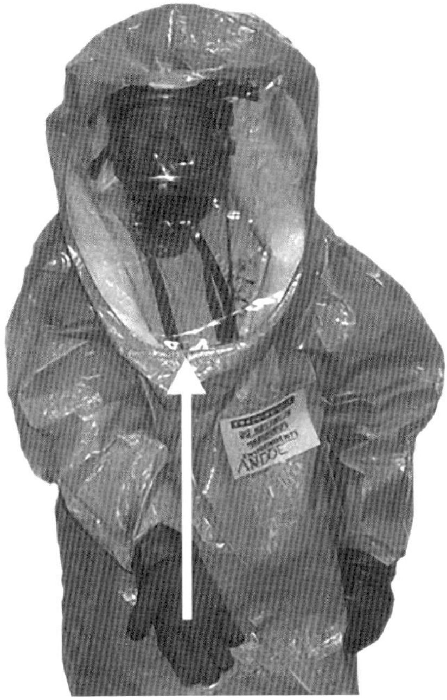

Figure 5-25

Follow me: raise open hand palm side facing direction in which travel is going to be made, and then lower hand towards ground.

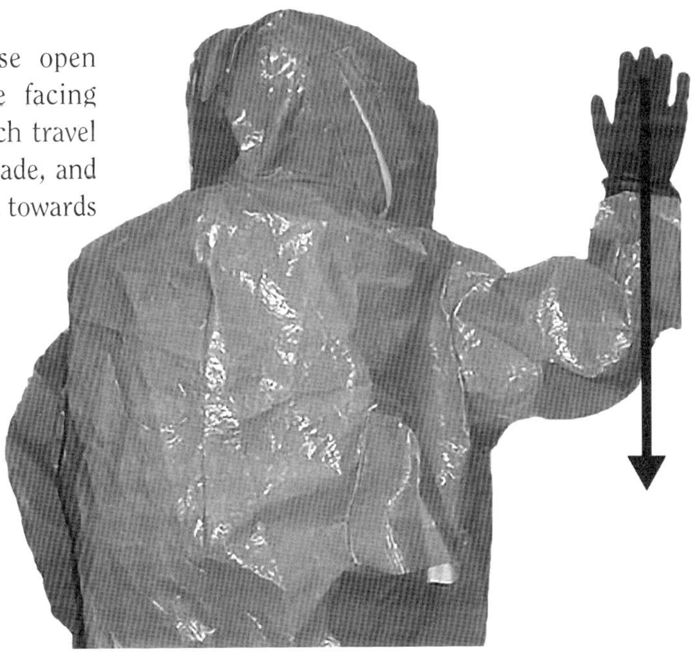

Figure 5-26

SECTION 6

Medical Considerations

Medical monitoring is an essential aspect to conducting operations in the hot zone. All emergency response teams should designate a medical professional to conduct pre and post entry medical monitoring for anyone who is planning to perform tasks in a hazardous environment where some level of PPE will be worn. This is especially true for personnel where SCBA in a level-A fully encapsulated chemical suit is required. Recording keeping is critical and considered a document subject to review under OSHA regulations. Additionally, all medical monitoring prior to and after response action in a designated hot zone should be annotated in the on-scene site safety health plan. Some other important information that should be found under the medical section of the site safety health plan include:

1. **Medical Aid Stations:** Should be indicated on site sketch identifying the location for monitoring, treatment, and rehabilitation of entry personnel, decontamination team and others as needed.
2. **Emergency Procedures:** These procedures will be used by on-site personnel. The safety officer must be notified of any on-site emergencies, and will be responsible for ensuring that the appropriate procedures are followed.
3. **Buddy Team:** Hot zone operations will be conducted using at least two-man teams. **NOTE:** Team members *will maintain eye contact at all times* to ensure location.
4. **Backup Team:** Primarily designated for emergencies. Located near the decontamination line and ready to conduct extraction of man-down once entry team is officially "on-air."

CAUTION: All emergency response personnel making entry in a designated hot zone in PPE should have pre and post entry medical vital signs taken by a qualified medically trained emergency medical technician. The designated medical technician should ensure proper record keeping. At a minimum, the completion of baseline medical vitals should be annotated in an individual's medical record prior to the responder conducting operations in the hot zone.

SECTION 7

Hot Zone Emergency Procedures

Given the hazardous environment responders are likely to encounter in a hot zone, it is important to have an established set of emergency protocols and procedures in place to safeguard personnel in the event of an emergency. This could be as basic as obtaining accurate readings from detection monitoring equipment to determine if the environment is volatile. Additionally, other significant reportable actions or information must be recognized and practiced to enhance the survivability of responders. The following is an example of a sequence of reportable information that can enhance existing organizational emergency protocols and procedures:

A. **Detection/monitoring equipment readings:**
   1. Radiation readings increase 2milli-rem (mR) above background – (est. 2mR line)
   2. Reading above 5 Rem/hr (egress except for lifesaving - 25 Rem/hr)
   3. LEL reading of 10%
   4. Oxygen levels above 22% or below 19%
   5. Other elevated readings

B. **Abort criteria:** This refers to any situation as seen by the responders in the hot zone as being a significant safety hazard that could result in death or serious injury. For an emergency situation that may require response personnel to immediately evacuate out of the hot zone consider the following:
   1. Responder Down (Injury/PPE breach/equipment failure).
   2. Any team member becomes a safety risk.
   3. Locate a booby trap or improvised explosive devise (IED) that has not been cleared by bomb detection unit.
   4. Communications Loss (3–5 minutes without visual contact or radio traffic).
   5. Fire/Explosion
   6. In the event of emergency occurring in the hot zone, all other operations cease.
   7. Situation deteriorating or events appear to be happening out of sequence.

Response agencies should develop SOP for all emergencies to include injury, PPE Failure, Fire, secondary devises/IED, and/or Explosion in the hot zone resulting in an alert to all response agencies:

C. **Emergency signal actions at the incident site:** In the event there is a significant emergency that can affect the safety of response personnel at the incident site, a protocol should be established through the incident command post that will immediately announce or signal to all personnel the presence of a hazard. The following is one recommendation on a technique to signal on-scene personnel of a hazard:
   1. **Primary signal:** "*abort mission*" should be communicated verbally.
   2. **Secondary signal:** *a series of three single horn blasts* will be sounded.
   3. **Responder action:** All site personnel will assemble at the cold end of the decontamination line for accountability and emergency situation/hazard update.

CAUTION: Avoid radio communication in the event the hazard is a secondary devise/IED found in the staging area or incident command post. Electronic transmission from radios can set off secondary devises/IED.

D. **Decontamination Procedures:** Personnel and equipment leaving the hot zone will be thoroughly decontaminated at the decontamination line positioned at the edge of the hot zone leading into the warm zone. At a minimum the makeup of an effective decontamination line should include:
   1. Primary containment – heavy-duty plastic sheet to capture/contain spray and run-off
   2. Secondary containment – initial gross decontamination pit, pool, or bucket to capture abundance of product runoff of both water and material from PPE
   3. Final rinse – containment system that replicates secondary containment
   4. Scrub brushes with extensions
   5. Soap and water source

Recommendations for decontamination solution are influenced by the type of hazard encountered. Be sure to check with appropriate references and supervisor before selecting type of decontamination solution. The following is a list of a few types available:

- Warm soap and water mixture (personnel)
- Bleach - 2.65% (50/50 household bleach and water) (equipment)
- Dry decontamination product - (temperature and reactivity specific)
- Caustic soda
- Sodium hypochlorite
- Soda ash

SECTION 8

Guidelines for Placard Identification

The following are recommended protocols to remember regarding placards and labeling when conducting operations in the hot zone where there are visible placards and labels posted on containers, drums, packages, or vehicles:

1. Approach incident cautiously from upwind to a point from which you can safely identify and/or read the placards. This may require binoculars if the upwind approach does not facilitate a good line of sight of the unknown hazardous materials.
2. Match the placards with those that are displayed in the most recently published Emergency Response Guidebook.
3. Consult the numbered guide associated with the sample placard to locate more critical information about the hazardous product.
4. Remember that guides associated with the placards provide only the most significant risk and/or hazard information.
5. Asterisks (*) on an orange placard represent explosives "Compatibility Group".
6. Products that are highlighted have a minimal required evacuation distance noted in the "green" pages of the ERG.

NOTE: The **Emergency Response Guidebook** (ERG200) was developed jointly by the US Department of Transportation, Transport Canada, and the Secretariat of Communications and Transportation of Mexico (SCT) for use by firefighters, police, and other emergency services personnel who may be the first to arrive at the scene of a transportation incident involving a hazardous material. It is primarily a guide to aid first responders in (1) quickly identifying the specific or generic classification of the material(s) involved in the incident, and (2) protecting themselves and the general public during this initial response phase of the incident. The ERG is updated every three to four years to accommodate new products and technology.

It is incumbent upon hazardous materials technicians to obtain as much information and resources (minimum three resources) before making tactical decisions regarding hazard response and hazard mitigation. Using fundamental techniques and tactics associated with hazard mitigation as an effort to minimize the extent of hazard release and or exposure is the ultimate goal of hazard response. Response agencies must develop their own techniques and practices that enable HazMat personnel to conduct safe offensive operations in the hot zone. Of importance is the assembly of proper mitigation equipment necessary to complete specified and or potential tactical goals. If possible, photo identification achieved from first entry teams conducting initial size up of the hazards can provide invaluable information on the potential products, their container(s), and their environment. Proper interpretation of relevant HazMat placards, labels, and material safety data sheets is extremely important. If done effectively, initial size up of the hazards will certainly influence tactical plans for mitigation and the actions and tasks for follow-on entry teams.

Contacting Information and Technical Support

- **Emergency Response Guidebook**—Copies are made available free of charge to public emergency responders through the **state coordinator** (United States only) nearest you. In Canada, contact CANUTEC at **613-992-4624** or via Internet at canutec@tc.gc.ca for information.
- **Hazardous Materials Information Center** can be contacted on **1-800-HMR-4922** (**1-800-467-4922**) or **202-366-4488** for Washington DC residents Monday through Friday from 9:00 am to 5:00 pm (EST).
- **The Office of Hazardous Materials Initiatives and Training** has the primary responsibility for providing training publications in support of the Hazardous Material Regulation at **202-366-4900**.

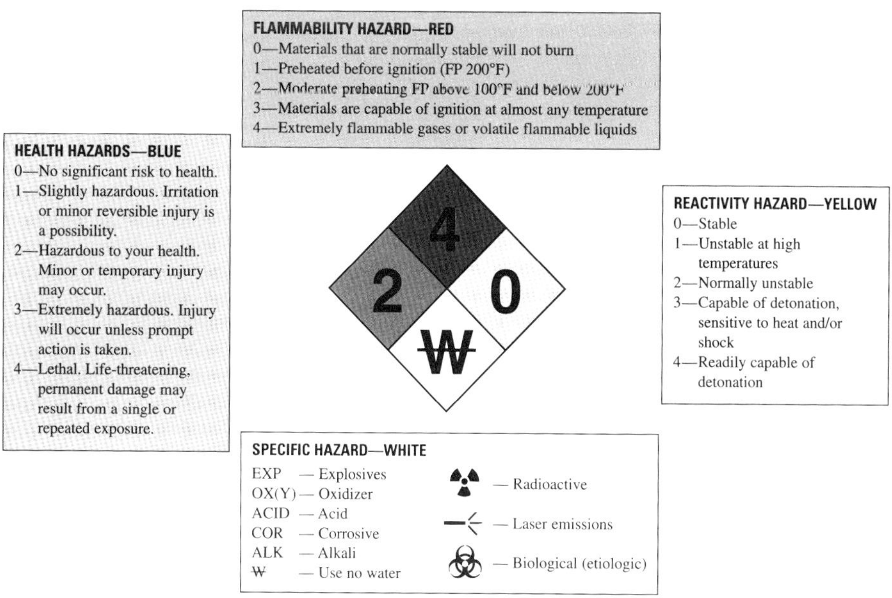

SECTION 9

Personal Training Record

This section serves as a personal record of the level of training and formal course work completed in the area of hazardous materials response and related subject matter courses. This section should be updated periodically as necessary to address most current required or supplemental training completed. Again, this log is for "YOU" the responder—it is incumbent upon each responder to maintain a current and accurate record of one's personal training status, records, credentials, and training performance as all of this material serves as a personal resume of response experience.

—SAMPLE PERSONAL TRAINING RECORD—

Course/Title: *HazMat Operating Site Practices*

Location: *Camp Creek County Fire Department Station*

Sponsored by: *National Fire Association*

Course Length/Hours Completed: *1 week - 40 hours*

Instructor's Name/Title: *Mr. Jim Argos - Dalton County Fire Captain*

Date Completed: *12 October 2002*

Course/Title: ______________________

Location: ______________________

Sponsored by: ______________________

Course Length/Hours Completed: ______________________

Instructor's Name/Title: ______________________

Date Completed: ______________________

Course/Title: ______________________________

Location: ______________________________

Sponsored by: ______________________________

Course Length/Hours Completed: ______________________________

Instructor's Name/Title: ______________________________

Date Completed: ______________________________

Course/Title: ______________________________

Location: ______________________________

Sponsored by: ______________________________

Course Length/Hours Completed: ______________________________

Instructor's Name/Title: ______________________________

Date Completed: ______________________________

Course/Title: ______________________________

Location: ______________________________

Sponsored by: ______________________________

Course Length/Hours Completed: ______________________________

Instructor's Name/Title: ______________________________

Date Completed: ______________________________

Course/Title: ______________________________

Location: ______________________________

Sponsored by: ______________________________

Course Length/Hours Completed: ______________________________

Instructor's Name/Title: ______________________________

Date Completed: ______________________________

SECTION 10

Personal Data Sheets

The Personal Data Sheets are a set of pre-formatted documents designed to capture individual actions in a hot zone and are the main focus of this Hot Zone Log. If used properly and consistently, these personal data sheets will provide an account of individual response actions that were performed in a hot zone. Additionally, from a health and safety perspective these personal data sheets will offer documented evidence as to the type, quantity, and concentration of hazards exposed to for a given response. The personal data sheets are most applicable for those responses or training drills where the use of level-A, level-B, SCBA, and or re-breather equipment is used. However, these data sheets can be used for any HazMat-related environment where a specified level of protection has been designated by a safety officer or training supervisors.

An example is provided of a sample personal data sheet that has been filled out based on a notional training mission. Please refer to previous sections or the glossary to assist in the recording of specific technical information that may enhance the data entry of each individual data sheet. It is recommended that the first line supervisor review the data in the log to ensure accuracy of response-specific information.

—SAMPLE PERSONAL DATA SHEET—

Hot zone entry # 23 Date: 9 Feb 2002

☐ Actual response ☒ Training

Site location: Washington Harbor, Middleton, TX

Site description: Commercial Cargo Ship Pier side with 140 containers

Incident name: SS Cargo Ship

Weather conditions: Partly cloudy, isolated showers

Temperature: 71 degrees

Other: Relative humidity 82%

Site Sketch:

Duty assignment: HazMat Specialist, Team Chief

Safety officer signature: Date: 29 June 2003

Hazard Evaluation:

(*refer to Section 11 regarding detection/monitoring equipment readings*)

Chemical: Organophosphate, corrosive products, hydrochloric acid cyanide gas

Biological: ______________________________

Radiological: ______________________________

Other: Shipboard structural hazards; electrical wire hazards second deck, port side______________________

Level of Protection PPE:

(*refer to Section 8 for PPE considerations*)

☒ Level A ☐ Level B ☐ Level C ☐ Level D

DECON Conducted:

☒ Yes ☐ No

Type of SCBA system used: Draeger-SCBA

Time on-air: 10:31 AM Time off-air: 10:54 AM

Total time on-air: 23 minutes

Starting air pressure: 4000 psi Ending: 2300 psi

Medical Monitoring Conducted:
(*refer to Section 10 for medical monitoring concerns*)

☒ Yes ☐ No

Pre-entry vital signs: Temp: 98.6° Pulse: 67 Resp: 14

B/P: 113/66 WT: 175 lbs. Other: no open cuts or sores

Post-entry vital signs: Temp: 97.2° Pulse: 74 Resp: 16

B/P: 111/78 WT: 173 Other:

Notes:

Job assignment was to work in two man crews to practice initial air monitoring and detection of gas leaking from container on second deck of SS Cargo ship. Initial pH reading indicated a corrosive environment with a pH of 3 located at the bottom right corner of a 40' container. Black and White Hazardous Materials Placard labeled Corrosive with number 8 was posted on the rear and side of container. Decision was made not to use detection equipment in corrosive environment as off-gassing of acidic product could damage sensitive monitoring equipment. Air sample and detection was achieved with Draeger kit and air collection tube. Visibility was good and there were few structural hazards other than loose electrical wire on second deck port side making for a trip hazard. Cargo ship carrying many chemical supplies. Ships manifest indicated the presence of hydrochloric acid in container where I received a pH of 3. Only one entry was made during this training drill.

—LOG SET 1 of 27—

Hot zone entry # ____________________ Date: ____________________

☐ Actual response ☐ Training

Site location: ____________________

Site description: ____________________

Incident name: ____________________

Weather conditions: ____________________

Temperature: ____________________

Other: ____________________

Site Sketch:

Duty assignment: ____________________

Safety officer signature: ____________________ Date __________

Hazard Evaluation:

Chemical: ____________________

Biological: ____________________

Radiological: ____________________

Other: ____________________

Level of Protection PPE:

☐ Level A ☐ Level B ☐ Level C ☐ Level D

DECON Conducted:

☐ Yes ☐ No

Type of SCBA system used: ____________________

Time on-air: ____________ Time off-air: ____________

Total time on-air: ____________________

Starting air pressure: ____________ Ending: ____________

Medical Monitoring Conducted:

☐ Yes ☐ No

Pre-entry vital signs: Temp: ________ Pulse: ________ Resp: ________

B/P: ________ WT: ________ Other: ________

Post-entry vital signs: Temp: ________ Pulse: ________ Resp: ________

B/P: ________ WT: ________ Other: ________

Notes:

—LOG SET 2 of 27—

Hot zone entry # ______________________ Date: ______________________

☐ Actual response ☐ Training

Site location: ______________________

Site description: ______________________

Incident name: ______________________

Weather conditions: ______________________

Temperature: ______________________

Other: ______________________

Site Sketch:

Duty assignment: ______________________

Safety officer signature: ______________________ Date ____________

Hazard Evaluation:

Chemical: ______________________

Biological: ______________________

Radiological: ______________________

Other: ______________________

Level of Protection PPE:

☐ Level A ☐ Level B ☐ Level C ☐ Level D

DECON Conducted:

☐ Yes ☐ No

Type of SCBA system used: ______________________________

Time on-air: ____________________ Time off-air: ____________________

Total time on-air: ______________________________

Starting air pressure: __________________ Ending: __________________

Medical Monitoring Conducted:

☐ Yes ☐ No

Pre-entry vital signs: Temp: ________ Pulse: ________ Resp: ________

B/P: ________ WT: ________ Other: ________

Post-entry vital signs: Temp: ________ Pulse: ________ Resp: ________

B/P: ________ WT: ________ Other: ________

Notes:

____

____

____

____

____

____

____

____

____

____

____

____

____

____

—LOG SET 3 of 27—

Hot zone entry # ______________________ Date: ______________________

☐ Actual response ☐ Training

Site location: ______________________

Site description: ______________________

Incident name: ______________________

Weather conditions: ______________________

Temperature: ______________________

Other: ______________________

Site Sketch:

Duty assignment: ______________________

Safety officer signature: ______________________ Date ____________

Hazard Evaluation:

Chemical: ______________________

Biological: ______________________

Radiological: ______________________

Other: ______________________

Level of Protection PPE:

☐ Level A ☐ Level B ☐ Level C ☐ Level D

DECON Conducted:

☐ Yes ☐ No

Type of SCBA system used: ______________________________

Time on-air: ______________ Time off-air: ______________

Total time on-air: ______________________________

Starting air pressure: ______________ Ending: ______________

Medical Monitoring Conducted:

☐ Yes ☐ No

Pre-entry vital signs: Temp: ________ Pulse: ________ Resp: ________

B/P: ________ WT: ________ Other: ________

Post-entry vital signs: Temp: ________ Pulse: ________ Resp: ________

B/P: ________ WT: ________ Other: ________

Notes:

____
____
____
____
____
____
____
____
____
____
____
____
____
____

—LOG SET 4 of 27—

Hot zone entry # ____________________ Date: ____________________

☐ Actual response ☐ Training

Site location: ____________________

Site description: ____________________

Incident name: ____________________

Weather conditions: ____________________

Temperature: ____________________

Other: ____________________

Site Sketch:

Duty assignment: ____________________

Safety officer signature: ____________________ Date ____________

Hazard Evaluation:

Chemical: ____________________

Biological: ____________________

Radiological: ____________________

Other: ____________________

Level of Protection PPE:

☐ Level A ☐ Level B ☐ Level C ☐ Level D

DECON Conducted:

☐ Yes ☐ No

Type of SCBA system used: ______________________________

Time on-air: ________________ Time off-air: ________________

Total time on-air: ______________________________

Starting air pressure: ______________ Ending: ______________

Medical Monitoring Conducted:

☐ Yes ☐ No

Pre-entry vital signs: Temp: ________ Pulse: ________ Resp: ________

B/P: ________ WT: ________ Other: ________

Post-entry vital signs: Temp: ________ Pulse: ________ Resp: ________

B/P: ________ WT: ________ Other: ________

Notes:

____

____

____

____

____

____

____

____

____

____

____

____

____

____

—LOG SET 5 of 27—

Hot zone entry # ____________________ Date: ____________________

☐ Actual response ☐ Training

Site location: ____________________

Site description: ____________________

Incident name: ____________________

Weather conditions: ____________________

Temperature: ____________________

Other: ____________________

Site Sketch:

Duty assignment: ____________________

Safety officer signature: ____________________ Date __________

Hazard Evaluation:

Chemical: ____________________

Biological: ____________________

Radiological: ____________________

Other: ____________________

Level of Protection PPE:

☐ Level A ☐ Level B ☐ Level C ☐ Level D

DECON Conducted:

☐ Yes ☐ No

Type of SCBA system used: ____________________

Time on-air: ____________ Time off-air: ____________

Total time on-air: ____________________

Starting air pressure: ____________ Ending: ____________

Medical Monitoring Conducted:

☐ Yes ☐ No

Pre-entry vital signs: Temp: ________ Pulse: ________ Resp: ________

B/P: ________ WT: ________ Other: ________

Post-entry vital signs: Temp: ________ Pulse: ________ Resp: ________

B/P: ________ WT: ________ Other: ________

Notes:

—LOG SET 6 of 27—

Hot zone entry # ____________________ Date: ____________________

☐ Actual response ☐ Training

Site location: ______________________________

Site description: ______________________________

Incident name: ______________________________

Weather conditions: ______________________________

Temperature: ______________________________

Other: ______________________________

Site Sketch:

Duty assignment: ______________________________

Safety officer signature: ____________________ Date __________

Hazard Evaluation:

Chemical: ______________________________

Biological: ______________________________

Radiological: ______________________________

Other: ______________________________

Level of Protection PPE:

☐ Level A ☐ Level B ☐ Level C ☐ Level D

DECON Conducted:

☐ Yes ☐ No

Type of SCBA system used: ______________________________

Time on-air: ________________ Time off-air: ________________

Total time on-air: ______________________________

Starting air pressure: ______________ Ending: ______________

Medical Monitoring Conducted:

☐ Yes ☐ No

Pre-entry vital signs: Temp: ________ Pulse: ________ Resp: ________

B/P: ________ WT: ________ Other: ________

Post-entry vital signs: Temp: ________ Pulse: ________ Resp: ________

B/P: ________ WT: ________ Other: ________

Notes:

____

____

____

____

____

____

____

____

____

____

____

____

____

____

—LOG SET 7 of 27—

Hot zone entry # ____________________ Date: ____________________

☐ Actual response ☐ Training

Site location: ____________________

Site description: ____________________

Incident name: ____________________

Weather conditions: ____________________

Temperature: ____________________

Other: ____________________

Site Sketch:

Duty assignment: ____________________

Safety officer signature: ____________________ Date __________

Hazard Evaluation:

Chemical: ____________________

Biological: ____________________

Radiological: ____________________

Other: ____________________

Level of Protection PPE:

☐ Level A ☐ Level B ☐ Level C ☐ Level D

DECON Conducted:

☐ Yes ☐ No

Type of SCBA system used: ______________________________

Time on-air: ________________ Time off-air: ________________

Total time on-air: ______________________________

Starting air pressure: ______________ Ending: ______________

Medical Monitoring Conducted:

☐ Yes ☐ No

Pre-entry vital signs: Temp: ________ Pulse: ________ Resp: ________

B/P: ________ WT: ________ Other: ________

Post-entry vital signs: Temp: ________ Pulse: ________ Resp: ________

B/P: ________ WT: ________ Other: ________

Notes:

____

____

____

____

____

____

____

____

____

____

____

____

____

____

—LOG SET 8 of 27—

Hot zone entry # ______________________ Date: ______________________

☐ Actual response ☐ Training

Site location: ______________________

Site description: ______________________

Incident name: ______________________

Weather conditions: ______________________

Temperature: ______________________

Other: ______________________

Site Sketch:

Duty assignment: ______________________

Safety officer signature: ______________________ Date ____________

Hazard Evaluation:

Chemical: ______________________

Biological: ______________________

Radiological: ______________________

Other: ______________________

Level of Protection PPE:

☐ Level A ☐ Level B ☐ Level C ☐ Level D

DECON Conducted:

☐ Yes ☐ No

Type of SCBA system used: ______________________

Time on-air: ______________ Time off-air: ______________

Total time on-air: ______________________

Starting air pressure: ______________ Ending: ______________

Medical Monitoring Conducted:

☐ Yes ☐ No

Pre-entry vital signs: Temp: ________ Pulse: ________ Resp: ________

B/P: ________ WT: ________ Other: ________

Post-entry vital signs: Temp: ________ Pulse: ________ Resp: ________

B/P: ________ WT: ________ Other: ________

Notes:

____

____

____

____

____

____

____

____

____

____

____

____

____

____

—LOG SET 9 of 27—

Hot zone entry # ____________________ Date: ____________________

☐ Actual response ☐ Training

Site location: ____________________

Site description: ____________________

Incident name: ____________________

Weather conditions: ____________________

Temperature: ____________________

Other: ____________________

Site Sketch:

Duty assignment: ____________________

Safety officer signature: ____________________ Date ____________

Hazard Evaluation:

Chemical: ____________________

Biological: ____________________

Radiological: ____________________

Other: ____________________

Level of Protection PPE:

☐ Level A ☐ Level B ☐ Level C ☐ Level D

DECON Conducted:

☐ Yes ☐ No

Type of SCBA system used: ______________________________

Time on-air: ________________ Time off-air: ________________

Total time on-air: ______________________________

Starting air pressure: ________________ Ending: ________________

Medical Monitoring Conducted:

☐ Yes ☐ No

Pre-entry vital signs: Temp: ________ Pulse: ________ Resp: ________

B/P: ________ WT: ________ Other: ________

Post-entry vital signs: Temp: ________ Pulse: ________ Resp: ________

B/P: ________ WT: ________ Other: ________

Notes:

____

____

____

____

____

____

____

____

____

____

____

____

____

____

—LOG SET 10 of 27—

Hot zone entry # ____________________ Date: ____________________

☐ Actual response ☐ Training

Site location: ____________________

Site description: ____________________

Incident name: ____________________

Weather conditions: ____________________

Temperature: ____________________

Other: ____________________

Site Sketch:

Duty assignment: ____________________

Safety officer signature: ____________________ Date ____________

Hazard Evaluation:

Chemical: ____________________

Biological: ____________________

Radiological: ____________________

Other: ____________________

Level of Protection PPE:

☐ Level A ☐ Level B ☐ Level C ☐ Level D

DECON Conducted:

☐ Yes ☐ No

Type of SCBA system used: ____________________

Time on-air: ____________ Time off-air: ____________

Total time on-air: ____________________

Starting air pressure: ____________ Ending: ____________

Medical Monitoring Conducted:

☐ Yes ☐ No

Pre-entry vital signs: Temp: ________ Pulse: ________ Resp: ________

B/P: ________ WT: ________ Other: ________

Post-entry vital signs: Temp: ________ Pulse: ________ Resp: ________

B/P: ________ WT: ________ Other: ________

Notes:

____

____

____

____

____

____

____

____

____

____

____

____

____

____

—LOG SET 11 of 27—

Hot zone entry # ______________________ Date: ______________________

☐ Actual response ☐ Training

Site location: ______________________

Site description: ______________________

Incident name: ______________________

Weather conditions: ______________________

Temperature: ______________________

Other: ______________________

Site Sketch:

Duty assignment: ______________________

Safety officer signature: ______________________ Date __________

Hazard Evaluation:

Chemical: ______________________

Biological: ______________________

Radiological: ______________________

Other: ______________________

Level of Protection PPE:

☐ Level A ☐ Level B ☐ Level C ☐ Level D

DECON Conducted:

☐ Yes ☐ No

Type of SCBA system used: ____________________

Time on-air: ____________ Time off-air: ____________

Total time on-air: ____________________

Starting air pressure: ____________ Ending: ____________

Medical Monitoring Conducted:

☐ Yes ☐ No

Pre-entry vital signs: Temp: ________ Pulse: ________ Resp: ________

B/P: ________ WT: ________ Other: ________

Post-entry vital signs: Temp: ________ Pulse: ________ Resp: ________

B/P: ________ WT: ________ Other: ________

Notes:

____

____

____

____

____

____

____

____

____

____

____

____

____

____

—LOG SET 12 of 27—

Hot zone entry # ____________________ Date: ____________________

☐ Actual response ☐ Training

Site location: ____________________

Site description: ____________________

Incident name: ____________________

Weather conditions: ____________________

Temperature: ____________________

Other: ____________________

Site Sketch:

Duty assignment: ____________________

Safety officer signature: ____________________ Date ____________

Hazard Evaluation:

Chemical: ____________________

Biological: ____________________

Radiological: ____________________

Other: ____________________

Level of Protection PPE:

☐ Level A ☐ Level B ☐ Level C ☐ Level D

DECON Conducted:

☐ Yes ☐ No

Type of SCBA system used: ____________________

Time on-air: ____________ Time off-air: ____________

Total time on-air: ____________________

Starting air pressure: ____________ Ending: ____________

Medical Monitoring Conducted:

☐ Yes ☐ No

Pre-entry vital signs: Temp: ________ Pulse: ________ Resp: ________

B/P: ________ WT: ________ Other: ________

Post-entry vital signs: Temp: ________ Pulse: ________ Resp: ________

B/P: ________ WT: ________ Other: ________

Notes:

—LOG SET 13 of 27—

Hot zone entry # ____________________ Date: ____________________

☐ Actual response ☐ Training

Site location: ____________________

Site description: ____________________

Incident name: ____________________

Weather conditions: ____________________

Temperature: ____________________

Other: ____________________

Site Sketch:

Duty assignment: ____________________

Safety officer signature: ____________________ Date __________

Hazard Evaluation:

Chemical: ____________________

Biological: ____________________

Radiological: ____________________

Other: ____________________

Level of Protection PPE:

☐ Level A ☐ Level B ☐ Level C ☐ Level D

DECON Conducted:

☐ Yes ☐ No

Type of SCBA system used: ______________________________

Time on-air: ______________ Time off-air: ______________

Total time on-air: ______________________________

Starting air pressure: ______________ Ending: ______________

Medical Monitoring Conducted:

☐ Yes ☐ No

Pre-entry vital signs: Temp: ________ Pulse: ________ Resp: ________

B/P: ________ WT: ________ Other: ________

Post-entry vital signs: Temp: ________ Pulse: ________ Resp: ________

B/P: ________ WT: ________ Other: ________

Notes:

____

____

____

____

____

____

____

____

____

____

____

____

____

____

—LOG SET 14 of 27—

Hot zone entry # ____________________ Date: ____________________

☐ Actual response ☐ Training

Site location: ____________________

Site description: ____________________

Incident name: ____________________

Weather conditions: ____________________

Temperature: ____________________

Other: ____________________

Site Sketch:

Duty assignment: ____________________

Safety officer signature: ____________________ Date __________

Hazard Evaluation:

Chemical: ____________________

Biological: ____________________

Radiological: ____________________

Other: ____________________

Level of Protection PPE:

☐ Level A ☐ Level B ☐ Level C ☐ Level D

DECON Conducted:

☐ Yes ☐ No

Type of SCBA system used: ______________________________

Time on-air: ______________ Time off-air: ______________

Total time on-air: ______________________________

Starting air pressure: ______________ Ending: ______________

Medical Monitoring Conducted:

☐ Yes ☐ No

Pre-entry vital signs: Temp: ________ Pulse: ________ Resp: ________

B/P: ________ WT: ________ Other: ________

Post-entry vital signs: Temp: ________ Pulse: ________ Resp: ________

B/P: ________ WT: ________ Other: ________

Notes:

____
____
____
____
____
____
____
____
____
____
____
____
____
____

—LOG SET 15 of 27—

Hot zone entry # ______________________ Date: ______________________

☐ Actual response ☐ Training

Site location: ______________________

Site description: ______________________

Incident name: ______________________

Weather conditions: ______________________

Temperature: ______________________

Other: ______________________

Site Sketch:

Duty assignment: ______________________

Safety officer signature: ______________________ Date ____________

Hazard Evaluation:

Chemical: ______________________

Biological: ______________________

Radiological: ______________________

Other: ______________________

Level of Protection PPE:

☐ Level A ☐ Level B ☐ Level C ☐ Level D

DECON Conducted:

☐ Yes ☐ No

Type of SCBA system used: ______________________________

Time on-air: ____________ Time off-air: ____________

Total time on-air: ______________________________

Starting air pressure: ____________ Ending: ____________

Medical Monitoring Conducted:

☐ Yes ☐ No

Pre-entry vital signs: Temp: ________ Pulse: ________ Resp: ________

B/P: ________ WT: ________ Other: ________

Post-entry vital signs: Temp: ________ Pulse: ________ Resp: ________

B/P: ________ WT: ________ Other: ________

Notes:

____

____

____

____

____

____

____

____

____

____

____

____

____

____

—LOG SET 16 of 27—

Hot zone entry # ______________________ Date: ______________________

☐ Actual response ☐ Training

Site location: ______________________

Site description: ______________________

Incident name: ______________________

Weather conditions: ______________________

Temperature: ______________________

Other: ______________________

Site Sketch:

Duty assignment: ______________________

Safety officer signature: ______________________ Date ____________

Hazard Evaluation:

Chemical: ______________________

Biological: ______________________

Radiological: ______________________

Other: ______________________

Level of Protection PPE:

☐ Level A ☐ Level B ☐ Level C ☐ Level D

DECON Conducted:

☐ Yes ☐ No

Type of SCBA system used: ____________________

Time on-air: ____________ Time off-air: ____________

Total time on-air: ____________________

Starting air pressure: ____________ Ending: ____________

Medical Monitoring Conducted:

☐ Yes ☐ No

Pre-entry vital signs: Temp: ________ Pulse: ________ Resp: ________

B/P: ________ WT: ________ Other: ________

Post-entry vital signs: Temp: ________ Pulse: ________ Resp: ________

B/P: ________ WT: ________ Other: ________

Notes:

____

____

____

____

____

____

____

____

____

____

____

____

____

____

—LOG SET 17 of 27—

Hot zone entry # ____________________ Date: ____________________

☐ Actual response ☐ Training

Site location: ____________________

Site description: ____________________

Incident name: ____________________

Weather conditions: ____________________

Temperature: ____________________

Other: ____________________

Site Sketch:

Duty assignment: ____________________

Safety officer signature: ____________________ Date ________

Hazard Evaluation:

Chemical: ____________________

Biological: ____________________

Radiological: ____________________

Other: ____________________

Level of Protection PPE:

☐ Level A ☐ Level B ☐ Level C ☐ Level D

DECON Conducted:

☐ Yes ☐ No

Type of SCBA system used: ______________________________

Time on-air: ____________________ Time off-air: ____________________

Total time on-air: ______________________________

Starting air pressure: __________________ Ending: __________________

Medical Monitoring Conducted:

☐ Yes ☐ No

Pre-entry vital signs: Temp: ________ Pulse: ________ Resp: ________

B/P: ________ WT: ________ Other: ________

Post-entry vital signs: Temp: ________ Pulse: ________ Resp: ________

B/P: ________ WT: ________ Other: ________

Notes:

____

____

____

____

____

____

____

____

____

____

____

____

____

____

—LOG SET 18 of 27—

Hot zone entry # ____________________ Date: ____________________

☐ Actual response ☐ Training

Site location: ____________________

Site description: ____________________

Incident name: ____________________

Weather conditions: ____________________

Temperature: ____________________

Other: ____________________

Site Sketch:

Duty assignment: ____________________

Safety officer signature: ____________________ Date __________

Hazard Evaluation:

Chemical: ____________________

Biological: ____________________

Radiological: ____________________

Other: ____________________

Level of Protection PPE:

☐ Level A ☐ Level B ☐ Level C ☐ Level D

DECON Conducted:

☐ Yes ☐ No

Type of SCBA system used: ____________________

Time on-air: ____________ Time off-air: ____________

Total time on-air: ____________________

Starting air pressure: ____________ Ending: ____________

Medical Monitoring Conducted:

☐ Yes ☐ No

Pre-entry vital signs: Temp: ________ Pulse: ________ Resp: ________

B/P: ________ WT: ________ Other: ________

Post-entry vital signs: Temp: ________ Pulse: ________ Resp: ________

B/P: ________ WT: ________ Other: ________

Notes:

—LOG SET 19 of 27—

Hot zone entry # ____________________ Date: ____________________

☐ Actual response ☐ Training

Site location: ____________________

Site description: ____________________

Incident name: ____________________

Weather conditions: ____________________

Temperature: ____________________

Other: ____________________

Site Sketch:

Duty assignment: ____________________

Safety officer signature: ____________________ Date __________

Hazard Evaluation:

Chemical: ____________________

Biological: ____________________

Radiological: ____________________

Other: ____________________

Level of Protection PPE:

☐ Level A ☐ Level B ☐ Level C ☐ Level D

DECON Conducted:

☐ Yes ☐ No

Type of SCBA system used: ______________________

Time on-air: ______________ Time off-air: ______________

Total time on-air: ______________________

Starting air pressure: ______________ Ending: ______________

Medical Monitoring Conducted:

☐ Yes ☐ No

Pre-entry vital signs: Temp: ________ Pulse: ________ Resp: ________

B/P: ________ WT: ________ Other: ________

Post-entry vital signs: Temp: ________ Pulse: ________ Resp: ________

B/P: ________ WT: ________ Other: ________

Notes:

____

____

____

____

____

____

____

____

____

____

____

____

____

____

—LOG SET 20 of 27—

Hot zone entry # ____________________ Date: ____________________

☐ Actual response ☐ Training

Site location: ____________________

Site description: ____________________

Incident name: ____________________

Weather conditions: ____________________

Temperature: ____________________

Other: ____________________

Site Sketch:

Duty assignment: ____________________

Safety officer signature: ____________________ Date __________

Hazard Evaluation:

Chemical: ____________________

Biological: ____________________

Radiological: ____________________

Other: ____________________

Level of Protection PPE:

☐ Level A ☐ Level B ☐ Level C ☐ Level D

DECON Conducted:

☐ Yes ☐ No

Type of SCBA system used: ______________________________

Time on-air: ________________ Time off-air: ________________

Total time on-air: ______________________________

Starting air pressure: ______________ Ending: ______________

Medical Monitoring Conducted:

☐ Yes ☐ No

Pre-entry vital signs: Temp: ________ Pulse: ________ Resp: ________

B/P: ________ WT: ________ Other: ________

Post-entry vital signs: Temp: ________ Pulse: ________ Resp: ________

B/P: ________ WT: ________ Other: ________

Notes:

____
____
____
____
____
____
____
____
____
____
____
____
____
____

—LOG SET 21 of 27—

Hot zone entry # ____________________ Date: ____________________

☐ Actual response ☐ Training

Site location: ____________________

Site description: ____________________

Incident name: ____________________

Weather conditions: ____________________

Temperature: ____________________

Other: ____________________

Site Sketch:

Duty assignment: ____________________

Safety officer signature: ____________________ Date __________

Hazard Evaluation:

Chemical: ____________________

Biological: ____________________

Radiological: ____________________

Other: ____________________

Level of Protection PPE:

☐ Level A ☐ Level B ☐ Level C ☐ Level D

DECON Conducted:

☐ Yes ☐ No

Type of SCBA system used: ______________________________

Time on-air: ________________ Time off-air: ________________

Total time on-air: ______________________________

Starting air pressure: ______________ Ending: ______________

Medical Monitoring Conducted:

☐ Yes ☐ No

Pre-entry vital signs: Temp: ________ Pulse: ________ Resp: ________

B/P: ________ WT: ________ Other: ________

Post-entry vital signs: Temp: ________ Pulse: ________ Resp: ________

B/P: ________ WT: ________ Other: ________

Notes:

____

____

____

____

____

____

____

____

____

____

____

____

____

____

—LOG SET 22 of 27—

Hot zone entry # ____________________ Date: ____________________

☐ Actual response ☐ Training

Site location: ____________________

Site description: ____________________

Incident name: ____________________

Weather conditions: ____________________

Temperature: ____________________

Other: ____________________

Site Sketch:

Duty assignment: ____________________

Safety officer signature: ____________________ Date __________

Hazard Evaluation:

Chemical: ____________________

Biological: ____________________

Radiological: ____________________

Other: ____________________

Level of Protection PPE:

☐ Level A ☐ Level B ☐ Level C ☐ Level D

DECON Conducted:

☐ Yes ☐ No

Type of SCBA system used: ______________________________

Time on-air: ______________ Time off-air: ______________

Total time on-air: ______________________________

Starting air pressure: ______________ Ending: ______________

Medical Monitoring Conducted:

☐ Yes ☐ No

Pre-entry vital signs: Temp: ________ Pulse: ________ Resp: ________

B/P: ________ WT: ________ Other: ________

Post-entry vital signs: Temp: ________ Pulse: ________ Resp: ________

B/P: ________ WT: ________ Other: ________

Notes:

____

____

____

____

____

____

____

____

____

____

____

____

____

____

—LOG SET 23 of 27—

Hot zone entry # ____________________ Date: ____________________

☐ Actual response ☐ Training

Site location: ____________________

Site description: ____________________

Incident name: ____________________

Weather conditions: ____________________

Temperature: ____________________

Other: ____________________

Site Sketch:

Duty assignment: ____________________

Safety officer signature: ____________________ Date ____________

Hazard Evaluation:

Chemical: ____________________

Biological: ____________________

Radiological: ____________________

Other: ____________________

Level of Protection PPE:

☐ Level A ☐ Level B ☐ Level C ☐ Level D

DECON Conducted:

☐ Yes ☐ No

Type of SCBA system used: ______________________________

Time on-air: ____________________ Time off-air: ____________________

Total time on-air: ______________________________

Starting air pressure: ________________ Ending: ________________

Medical Monitoring Conducted:

☐ Yes ☐ No

Pre-entry vital signs: Temp: ________ Pulse: ________ Resp: ________

B/P: ________ WT: ________ Other: ________

Post-entry vital signs: Temp: ________ Pulse: ________ Resp: ________

B/P: ________ WT: ________ Other: ________

Notes:

____

____

____

____

____

____

____

____

____

____

____

____

____

____

—LOG SET 24 of 27—

Hot zone entry # ____________________ Date: ____________________

☐ Actual response ☐ Training

Site location: ____________________

Site description: ____________________

Incident name: ____________________

Weather conditions: ____________________

Temperature: ____________________

Other: ____________________

Site Sketch:

Duty assignment: ____________________

Safety officer signature: ____________________ Date __________

Hazard Evaluation:

Chemical: ____________________

Biological: ____________________

Radiological: ____________________

Other: ____________________

Level of Protection PPE:

☐ Level A ☐ Level B ☐ Level C ☐ Level D

DECON Conducted:

☐ Yes ☐ No

Type of SCBA system used: ____________________

Time on-air: __________ Time off-air: __________

Total time on-air: ____________________

Starting air pressure: __________ Ending: __________

Medical Monitoring Conducted:

☐ Yes ☐ No

Pre-entry vital signs: Temp: ________ Pulse: ________ Resp: ________

B/P: ________ WT: ________ Other: ________

Post-entry vital signs: Temp: ________ Pulse: ________ Resp: ________

B/P: ________ WT: ________ Other: ________

Notes:

—LOG SET 25 of 27—

Hot zone entry # ______________________ Date: ______________________

☐ Actual response ☐ Training

Site location: ______________________

Site description: ______________________

Incident name: ______________________

Weather conditions: ______________________

Temperature: ______________________

Other: ______________________

Site Sketch:

Duty assignment: ______________________

Safety officer signature: ______________________ Date ____________

Hazard Evaluation:

Chemical: ______________________

Biological: ______________________

Radiological: ______________________

Other: ______________________

Level of Protection PPE:

☐ Level A ☐ Level B ☐ Level C ☐ Level D

DECON Conducted:

☐ Yes ☐ No

Type of SCBA system used: ____________________

Time on-air: ____________ Time off-air: ____________

Total time on-air: ____________________

Starting air pressure: ____________ Ending: ____________

Medical Monitoring Conducted:

☐ Yes ☐ No

Pre-entry vital signs: Temp: ________ Pulse: ________ Resp: ________

B/P: ________ WT: ________ Other: ________

Post-entry vital signs: Temp: ________ Pulse: ________ Resp: ________

B/P: ________ WT: ________ Other: ________

Notes:

—LOG SET 26 of 27—

Hot zone entry # ______________________ Date: ______________________

☐ Actual response ☐ Training

Site location: ______________________

Site description: ______________________

Incident name: ______________________

Weather conditions: ______________________

Temperature: ______________________

Other: ______________________

Site Sketch:

Duty assignment: ______________________

Safety officer signature: ______________________ Date ____________

Hazard Evaluation:

Chemical: ______________________

Biological: ______________________

Radiological: ______________________

Other: ______________________

Level of Protection PPE:

☐ Level A ☐ Level B ☐ Level C ☐ Level D

DECON Conducted:

☐ Yes ☐ No

Type of SCBA system used: ______________________________

Time on-air: ______________ Time off-air: ______________

Total time on-air: ______________________________

Starting air pressure: ______________ Ending: ______________

Medical Monitoring Conducted:

☐ Yes ☐ No

Pre-entry vital signs: Temp: ________ Pulse: ________ Resp: ________

B/P: ________ WT: ________ Other: ________

Post-entry vital signs: Temp: ________ Pulse: ________ Resp: ________

B/P: ________ WT: ________ Other: ________

Notes:

____

____

____

____

____

____

____

____

____

____

____

____

____

____

—LOG SET 27 of 27—

Hot zone entry # ____________________ Date: ____________________

☐ Actual response ☐ Training

Site location: ____________________

Site description: ____________________

Incident name: ____________________

Weather conditions: ____________________

Temperature: ____________________

Other: ____________________

Site Sketch:

Duty assignment: ____________________

Safety officer signature: ____________________ Date __________

Hazard Evaluation:

Chemical: ____________________

Biological: ____________________

Radiological: ____________________

Other: ____________________

Level of Protection PPE:

☐ Level A ☐ Level B ☐ Level C ☐ Level D

DECON Conducted:

☐ Yes ☐ No

Type of SCBA system used: ______________________________

Time on-air: ________________ Time off-air: ________________

Total time on-air: ______________________________

Starting air pressure: ______________ Ending: ______________

Medical Monitoring Conducted:

☐ Yes ☐ No

Pre-entry vital signs: Temp: _______ Pulse: _______ Resp: _______

B/P: _______ WT: _______ Other: _______

Post-entry vital signs: Temp: _______ Pulse: _______ Resp: _______

B/P: _______ WT: _______ Other: _______

Notes:

SECTION 11

Common Terms and Definitions

Absorption materials may be taken into the body (absorbed) through the skin or lungs.

Acid any chemical with a low pH that can cause skin or tissue damage. (See also pH.)

Acute effect a health effect which develops rapidly. Exposure to carbon monoxide may cause a person to pass out, an acute effect

Administrative controls policies and practices written before work begins to minimize exposure to chemical and physical hazards.

Adsorption when a chemical becomes attached to another material, for example, soil.

Air-purifying respirator protective mask which filters air and toxic materials through filters.

Alkali any chemical with a high pH that in water solution is bitter and more or less irritating or caustic to the skin. Strong alkalies in solution are corrosive to the skin and mucous membranes. (See also pH.)

Alpha positively charged radiation particle capable of traveling only a few inches in air.

Alveoli the small air spaces deep in the lung where oxygen goes into the blood.

Anhydrous free from water.

Antagonism the situation in which two chemicals, when introduced, interfere with each other's actions, or when one chemical interferes with the action of the other.

Asphyxiant a vapor or gas which can cause unconsciousness or death by suffocation (lack of oxygen). Asphyxiation is one of the principal potential hazards of working in confined spaces.

Base a liquid or solid which has a pH higher than 7.

Beta a negatively charged radiation particle.

Boiling point temperature at which a liquid changes to a vapor. Flammable materials with low boiling points generally present special fire hazards.

Buddy system a work practice in which workers team up in pairs during work activities.

Bung the cap or plug that fits into the holes in a drum.

Carcinogen a substance which can cause cancer.

CAS Number a unique number assigned to a chemical by the Chemical Abstract Service.

Catalyst a substance that speeds up a chemical reaction.

cc cubic centimeter, a metric measurement about the size of a sugar cube.

Ceiling (C) the maximum allowable exposure limit for an airborne substance, not to be exceeded during the shift.

Central Nervous System (CNS) the part of the nervous system protected by the skull and the spinal column.

Chemical cartridge a filtering device which is attached to an air-purifying respirator.

Chronic effect a health effect which develops slowly over a long period of time.

Comprehensive Environmental Response, Compensation and Liability Act (CERCLA) authorized government money for clean-up of abandoned hazardous waste sites, clean-up and emergency response to transportation incidents involving chemical releases, payment to injured or diseased citizens, etc. It was amended in 1986.

Concentration the amount of one material in another.

Confined space any area which limits natural ventilation, making it easier for gases or vapors to accumulate.

Corrosive a liquid or solid that eats away another material or substance with which it comes in contact.

cu m (or m^3) cubic meter.

Decomposition the breakdown of a material (by heat, chemical reaction, decay, or other processes).

Decontamination the chemical or physical process of reducing and preventing the spread of contamination from persons and equipment used at a hazardous materials incident.

Decontamination line a line set up with stations for decontamination procedures between the Exclusion Zone and the Support Zone.

Degradation method by which the protective properties of CPC are diminished.

Demand regulator reduces tank pressure to provide air when the wearer inhales.

Department of Transportation (DOT) regulates shipments and transfer of hazardous materials.

Dermatitis redness or irritation of the skin.

Dilution method of reducing the concentration of a contaminant to a safe level.

Dose the quantity of a chemical taken into the body.

Dose response the relationship between the amount of the chemical and the amount of response in humans or animals.

Dose response curve a graph that shows how much of the chemical dose causes an observed effect.

Drum grappler a device designed to be attached to mechanical equipment for drum handling.

Emergency an unexpected or unplanned event requiring remedial action.

Emergency response plan written descriptions of planned actions and personnel responsibilities for emergency response actions.

Engineering control substitution, isolation, and ventilation methods used to reduce the level of the contaminant at the source.

Environmental Protection Agency (EPA) federal agency concerned with environmental air, water, and land quality.

Evaporation rate how fast a liquid enters the air when compared with a known material. The evaporation rate can be useful in evaluating the health and fire hazards of a material.

Exclusion zone (Hot Zone) the area where the hazard is present—contaminated area.

Exposure the concentration of a material in the air to which a worker can come into contact. Usually, exposure is measured near the nose of the worker.

Flammable a liquid with a flash point below 100°F (37.8°C)—Class I A: those having flash points below 73°F and having a boiling point below 100°F; Class I B: those having flash points below 73°F and having a boiling point at or above 100°F; Class I C: those having flash points at or above 73°F and below 100°F.

Flash point the temperature at which a liquid will give off enough flammable vapor to burn. There are several flash point test methods, and flash points may vary for the same material, depending on the method used; so the test method is indicated when the flash point is given.

Gram (gm) a metric unit of weight. One U.S. ounce is about 28.4 grams.

Hazardous material to be considered hazardous, a waste must be on the list of specific hazardous waste streams or chemicals, or else it must exhibit one or more of certain specific characteristics, including flammability, corrosivity, reactivity, and toxicity. The definition excludes household waste, agricultural waste returned to the soil, and mining overburden returned to the mine site. It also excludes all wastewater returned directly or indirectly to surface waters. However, hazardous waste may physically be in the liquid state.

Hazardous Waste Operations and Emergency Response (HAZWOPER) OSHA standard which was developed to protect hazardous waste personnel and emergency response personnel.

Hazards the inherent characteristics of a material that may cause incapacitation, injury, or mortality by contact, inhalation, or ingestion.

HAZCOM (1910.1200) OSHA Hazard Communication Standard.

Heat exhaustion prolonged exposure to intense heat exceeds the body's ability to cool down, causing excessive sweating and sodium deficiency.

Heat stroke a life-threatening condition during prolonged exposure to intense heat when the body is unable to sweat; extremely high body temperature and collapse may result.

Heavy metals the major toxic metals.

Hematotoxin toxic to the blood or organs where blood is made.

Hepatotoxin toxic to the liver

Immediately Dangerous to Life and Health (IDLH) any condition which may result in damage to health which cannot be repaired. IDLH situations include explosive and oxygen-deficient environments and the presence of Class A poisons or substances which can be absorbed through the skin.

Incident Command System (ICS) an organized system of personnel and delegation of responsibilities which controls the response to an emergency.

Incident Commander person in charge of on-site management of all activities at a hazardous materials emergency.

Incompatible chemicals chemicals which produce a negative reaction when mixed.

Ingestion taking a substance in through the mouth.

Inhalation breathing in through the mouth or nose a substance in the form of a gas, vapor, fume, mist, or dust.

Irritant a substance which causes an inflammatory response when brought into contact with the eyes, skin, or respiratory system.

Isolation method of decontamination in which contaminated equipment and materials are bagged or covered and set aside, usually for subsequent shipment to an approved landfill for disposal.

Latency the time interval between exposure to a substance and the development of a disease.

Leachate liquid released into soil from a land disposal facility. Leachate is generated when water enters a landfill, migrates through it, picking up soluble materials, and seeps into the soil.

Lock-out a procedure to prevent use of any equipment which should not be used, generally tagged and locked closed or off.

Lower explosive limit (LEL) the lowest concentration (lowest percentage of the substance in air) that will produce a flash of fire when an ignition source (heat, arc, or flame) is present.

M meter; a metric unit of length equal to about 39 inches.

Main-line valve controls air flow to the regulator on supplied-air respirators.

Manifest form required by the EPA to track hazardous wastes.

Material Safety Data Sheet (MSDS) informational sheet which is sent with hazardous materials. Lists chemical properties, emergency response procedures, reactivity data, control measures, safe handling procedures, and manufacturer.

Melting point the temperature at which a solid substance changes to a liquid state. For mixtures, a melting range may be given.

Metabolism the chemical reactions that go on in the body to maintain life.

Milligrams per cubic meter (mg/m^3) unit of measurement generally used to weigh fibers. One U.S. gram is about 1,000 milligrams.

Monitoring measuring concentrations of substances in the workplace.

Morbidity non-fatal disease from an exposure.

Mortality death from an exposure.

Mutagen a substance which can change the genetic material (DNA) in a living cell.

National Fire Protection Association (NFPA) produces many standards, including the four-color diamond used on labels to indicate hazard. Health, fire, and reactivity hazards are rated from 0 (none) to 4 (extreme). The Health rating is in the blue section, Fire in red, and Reactivity in yellow. The white section is reserved for Other Specific Hazards (for example, "radiation: do not use water" or "fire").

National Institute for Environmental Health Sciences (NIEHS) a federal agency responsible for issues related to the environment.

National Institute for Occupational Safety and Health (NIOSH) a federal research agency responsible for issues related to occupational safety and health.

Neurotoxin a substance which is toxic to the brain and nerves.

Nuclear Regulatory Commission responsible for community and worker protection from radiation hazards.

Occupational Safety and Health Administration (OSHA) a federal unit responsible for creating and enforcing laws related to occupational safety and health.

Olfactory relating to the nose or sense of smell.

Oral taken into the body through the mouth.

Oral toxicity adverse effects resulting from taking a substance into the body through the mouth.

Oxidation a reaction in which a substance combines with oxygen, the oxygen being provided by an oxidizer or oxidizing agent.

Oxidizer a substance that gives up oxygen readily to stimulate the combustion of organic matter.

Oxygen deficient air which contains less than 19.5% oxygen.

Oxygen enriched an atmosphere containing more than 22% oxygen.

Parts per million (ppm) a unit used to measure particulates.

Penetration the flow of a chemical through zippers, stitched seams, pores, or imperfections in the material.

Percutaneous absorption absorption into the skin.

Permeation process by which a chemical dissolves in or moves through a protective clothing material on a molecular level.

Permissible Exposure Limit (PEL) the highest level of a substance to which a person can be exposed, set by OSHA.

pH applies to liquids. A pH less than 7 is acid. A pH of 7 is neutral, and a pH greater than 7 is alkaline.

Physical agent heat, noise, radiation, vibration.

Pulmonary toxin toxic to the lungs.

Qualitative fit test measures effectiveness of a respirator by exposing wearer to a test atmosphere containing an irritating or smelly substance. Wearer should not be able to detect the substance.

Quantitative fit test measures effectiveness of a respirator in preventing substance from entering the facepiece while wearer is in a test chamber. Actual amount of concentration of substance is measured inside the facepiece of the respirator.

Rad A measure of radiation energy absorbed by the body.

Reactivity a description of the tendency of a substance to undergo chemical reaction with the release of energy. Undesirable effects such as pressure buildup, temperature increase, or formation of noxious, toxic, or corrosive by-products may occur because of the reactivity of a substance to heating, burning, direct contact with other material, or other conditions found in use or in storage.

REM Roentgen Equivalent Man, a measure of radiation dose.

Renal pertaining to the kidney.

Residual volume (RV) the amount of air remaining in the lung after maximum expiratory effort.

Resource Conservation and Recovery Act (RCRA) a federal law which regulates management and disposal of hazardous materials and wastes.

Risk exposure to the chance of injury or loss.

Route of entry how material gets into the body: inhaled, ingested, via the skin (dermal), or eye(s).

Self-Contained Breathing Apparatus (SCBA) a supplied-air respirator consisting of an oxygen tank, carrying assembly, gauge, safety valve, and full facepiece for use when exposures are unknown or particularly toxic.

Sensitizer a substance which on first exposure causes little or no reaction but which on repeated exposure may cause a marked response, not necessarily limited to the contact site.

Short-Term Exposure Limit (STEL) an exposure limit set by ACGIH that sets the maximum concentration that a worker can be exposed to during a 15-minute period.

Solubility in water a measure of how much of a material will dissolve in water.

Specific gravity the weight of a material compared to the weight of an equal volume of water; an expression of the density (or heaviness) of the material. Insoluble materials with specific gravity less than 1.0 will float in water; insoluble materials with specific gravity greater than 1.0 will sink in water.

Stability an expression of the ability of a material to remain unchanged. For MSDS purposes, a material is stable if it remains in the same form under expected and reasonable conditions of storage or use. Conditions which may cause instability (dangerous change) are stated. Examples are temperatures above 150°F or shock from dropping.

Standard Operating Procedures (SOP) written descriptions of tasks and activities to be followed during work.

Support Zone (Cold Zone) area where administrative and support functions not requiring respiratory protective equipment are performed.

Synergistic effect a combined effect of two or more substances which is greater than the sum of the effect of each.

Systemic relating to the whole body rather than its individual parts.

Teratogen a substance or agent to which exposure of a pregnant female can result in changes in the fetus.

Threshold the lowest dose or exposure to a chemical at which a specific effect is observed.

Threshold Limit Value (TLV) A term used by ACGIH to express the airborne concentration of a material to which nearly all persons can be exposed day after day without adverse effects. ACGIH express TLVs in three ways:

TLV-TWA the allowable time-weighted average concentration for a normal 8-hour work day.

TLV-STEL the short-term exposure limit, or maximum concentration for a continuous 15-minute exposure period (maximum of four such periods a day, with at least 60 minutes between exposure periods and provided that the daily TLV-TWA is not exceeded).

TLV-C The ceiling exposure limit or "the concentration that should not be exceeded during any part of the working exposure . . . a 15-minute period except for those substances which may cause immediate irritation . . ."

Time-Weighted Average (TWA) measurement to determine the worker's average exposure to a substance over a typical 8-hour work shift. The actual exposure is then compared to OSHA standards or other professional guidelines.

Toxic Substances Control Act (TSCA) a list of reviewed harmful substances that require precautions and safe work practices by the community and industry.

Toxicity the adverse effects from exposure.

United Nations' Identification Number (UN Number) a number used internationally to identify a hazardous material.

United States Coast Guard (USCG) concerned with the transportation of hazardous materials across navigable waterways and other bodies of water.

Upper Explosive Limit or Upper Flammable Limit (UEL/UFL) pertaining to a vapor or gas. The highest concentration (highest percentage of the substance in air) that will produce a flash of fire when an ignition source (heat, arc, or flame) is present. At higher concentrations, the mixture is too "rich" to burn. Also see "LEL."

Vapor gaseous form of a substance normally in the liquid or solid state at room temperature.

Vapor density the weight of a vapor or gas compared to the weight of an equal volume of air; an expression of the density of the vapor or gas. Materials lighter than air have vapor densities less than 1.0. Materials heavier than air have vapor densities greater than 1.0.

Vapor pressure indicates the tendency of a liquid to evaporate into the air.

Ventilation a form of engineering control that removes airborne contaminants.

Viscosity the property of liquid; resistance to flow.

Waste profile sheet a document which is provided by the laboratory that conducted the analysis of the hazardous waste. Describes the physical and chemical properties of the waste sample.

Source: Hazardous Materials Training Center Last Updated 11/26/2002© 2002 University of Louisville